ELECTRICITY

BY SALLY M WALKER
PHOTOGRAPHS BY ANDY KING

LERNER BOOKS • LONDON • NEW YORK • MINNEAPOLIS

This book was first published in the United States of America in 2006.

First published in the United Kingdom in 2008 by
Lerner Books,
Dalton House,
60 Windsor Avenue,
London SW19 2RR

This edition was updated and edited for UK publication by Discovery Books Ltd., Unit 3, 37 Watling Street, Leintwardine, Shropshire SY7 0LW

British Library Cataloguing in Publication Data

Walker, Sally M.
  Electricity. - (Early bird energy)
  1. Electricity - Juvenile literature
  I. Title
  537

  ISBN-13: 978 1 58013 310 4

Additional photographs with permission of: © The Discovery Picture Library, p. 4; © Photodisc/Getty Images, pp. 5, 28; © Royalty-Free/CORBIS, p. 8.

Printed in China

CONTENTS

Be A Word Detective 5

Chapter 1
PEOPLE AND ELECTRICITY 6

Chapter 2
AMAZING ATOMS 10

Chapter 3
ELECTRICAL CHARGE 16

Chapter 4
CURRENTS AND CIRCUITS 24

A Note to Adults on Sharing a Book 44

Learn More about
Electricity and Energy 45

Glossary 46

Index 48

BE A WORD DETECTIVE

Can you find these words as you read about electricity?
Be a detective and try to work out what they mean.
You can turn to the glossary on page 46 for help.

atoms	electrons	orbit
circuit	insulator	particles
conductors	ion	positive charge
current	negative charge	static electricity
electrical charge	nucleus	terminal

CHAPTER 1
PEOPLE AND ELECTRICITY

Long ago, people lit their homes with candles.
Today most people use electricity. Electricity is a
form of energy. Electrical energy can be used in
many ways.

6

Look around your classroom or home. What
electrical things can you see? Lights?
A computer? A television?

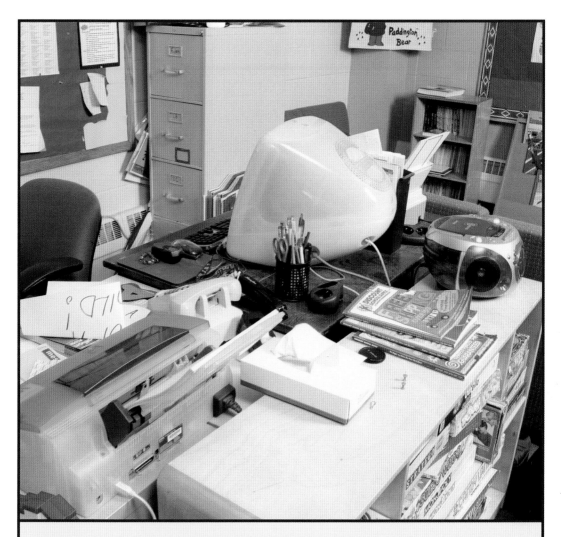

**Electricity is being used all around us, in everything from radios and
computers to traffic lights and street lights.**

Electricity is helpful. But it can be dangerous! So remember these safety rules. Do not touch electrical sockets. Never stick anything inside an electrical socket. Keep electrical wires and appliances far away from water. Do not touch broken or cracked electrical wires. Never try to break open a battery. Stay away from outdoor power lines. Go inside during a thunderstorm, because lightning is electricity.

Electricity is powerful. It can hurt or even kill people.

All scientists need to stay safe. Be sure you have an adult's help when you do experiments.

The experiments in this book are safe. But before doing them, talk with an adult. He or she might like to help you experiment.

How does electricity make a lamp light up?

CHAPTER 2
AMAZING ATOMS

When you turn on a lamp, electricity lights the bulb. But where does electricity begin? It begins inside tiny particles called atoms. Atoms are so small that you can't see them.

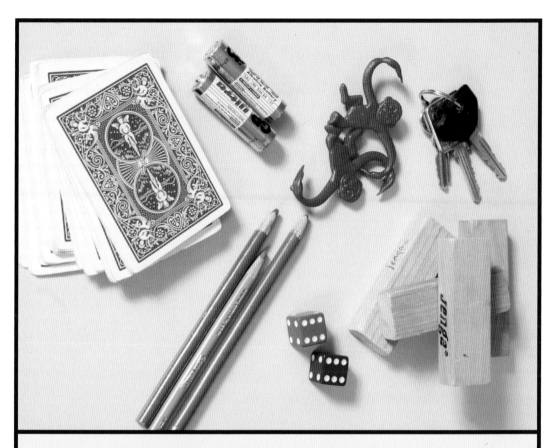

All of these things are made of atoms. Different kinds of atoms combine in different ways to make up everything we see.

Everything around you is made of atoms. There are more than 100 different kinds of atoms. Atoms can join together in many different ways. That's why we have different substances, like air, apples and toys.

An atom has three main kinds of parts. These parts are called protons, electrons and neutrons. Protons and electrons have electrical energy. A proton's electrical energy has a positive charge. An electron's electrical energy has a negative charge. Neutrons have no charge. The whole atom does not have a charge either. That's because the protons and the electrons balance each other out.

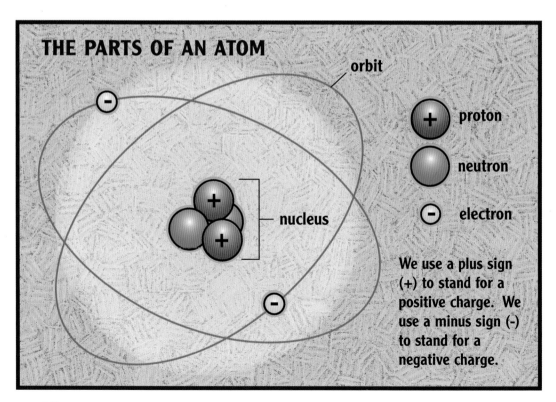

THE PARTS OF AN ATOM

orbit

proton

neutron

electron

nucleus

We use a plus sign (+) to stand for a positive charge. We use a minus sign (-) to stand for a negative charge.

These boys and girls are standing near each other. They are like protons and neutrons crowded close together in an atom's nucleus. If they are the nucleus, where would the electrons be in this picture?

Protons and neutrons are in an atom's centre. This centre is called the nucleus. Electrons circle around the nucleus. Their path is called an orbit. Some electrons circle close to the nucleus. Others follow orbits that are further away.

Rubbing objects together can move electrons around. Sometimes an electron gets knocked out of its orbit. Then it is called a free electron.

Did you know that when you pet a dog, you may be moving electrons around? Rubbing the animal's fur can move the electrons.

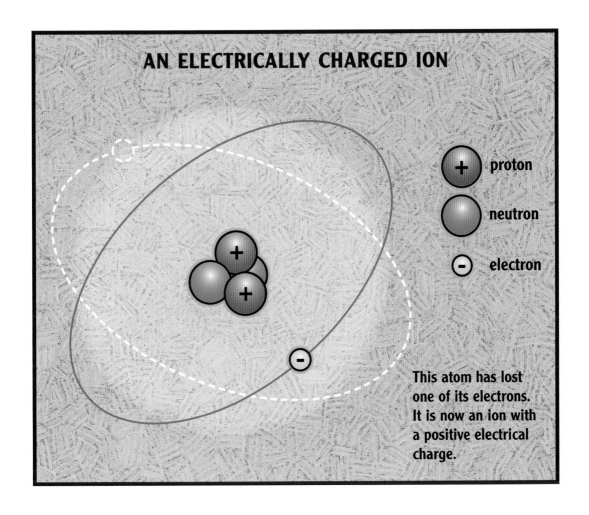

AN ELECTRICALLY CHARGED ION

proton

neutron

electron

This atom has lost one of its electrons. It is now an ion with a positive electrical charge.

A free electron may jump to another atom. When atoms gain or lose electrons, they become ions. An ion is an atom with an electrical charge. Ions with extra electrons have a negative charge. Ions with too few electrons have a positive charge.

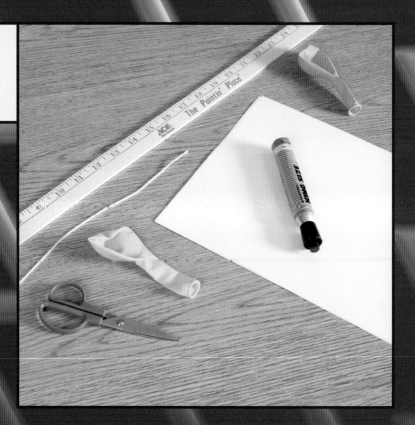

How can you use these simple things to make an electrical charge?

CHAPTER 3
ELECTRICAL CHARGE

Most objects have no electrical charge. An object becomes electrically charged when it loses or gains electrons. You can prove this yourself. You will need two balloons, a piece of paper, a 40-centimetre long piece of string, a metre rule, a marker pen and scissors.

Draw several penny-sized circles on the paper. Cut the circles out. Put them on a table. Draw a very small X on each balloon. Blow up both balloons and tie them shut.

Carefully cut out the paper circles.

Pick up one of the balloons. Rub the *X* against your hair 15 times. Rubbing makes some electrons in your hair leave their orbits. These free electrons move from your hair to the balloon. Now the spot marked with an *X* on the balloon has extra electrons. It has a negative charge.

There are millions of electrons in every one of your hairs. Rubbing the balloon against your hair moves those electrons.

Watch the paper circles closely.

Now hold the *X* on the balloon 4 centimetres above the paper circles. What happens? The circles jump onto the balloon. Why?

A kind of electricity called static electricity holds the paper circles to the balloon. Static electricity is energy created between two objects with different charges. Different charges are also called opposite charges. Opposite charges pull themselves towards each other.

Static electricity makes the paper stick to the balloon.

X marks the spot! Placing an *X* on your balloon will help you know what part of the balloon is charged.

The part of the balloon that you marked with an *X* has a negative charge. That negative charge pulls positive charges in the paper circles towards the balloon. Static electricity forms and holds the paper and the balloon together.

Two objects with the same kind of charge have like charges. Like charges push away from each other. Prove it! Tie one end of the string around the knot of one balloon. Tie the other end around one end of the metre rule.

Put the metre rule on a table so the balloon hangs down. Lift the balloon. Rub the marked *X* on your hair 15 times. Let go of the balloon.

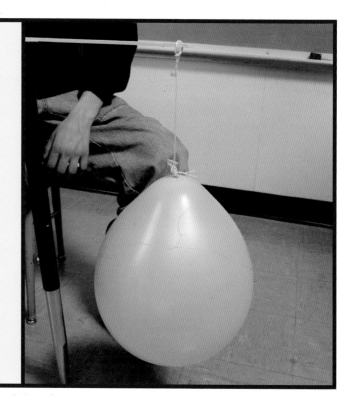

After rubbing the balloon against your hair, let it hang straight down on its string.

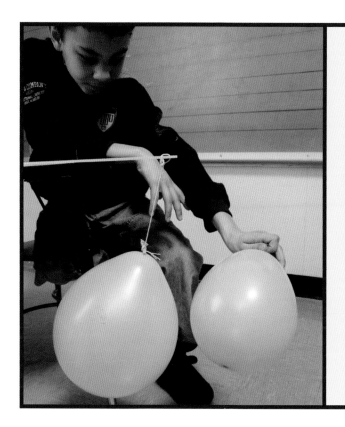

You can use static electricity to move the hanging balloon without even touching it.

Pick up the other balloon. Rub its X on your hair 15 times. Try to touch the X on this balloon to the X on the hanging balloon. (Don't touch the hanging balloon with your hand.) What happens? The X on the hanging balloon moves away from the balloon in your hand. Why? Both balloons have a negative charge. Like charges push away from each other.

Static electricity can make your hair stand up like this—but only for a little while. Why doesn't static electricity last very long?

CHAPTER 4
CURRENTS AND CIRCUITS

Static electricity lasts only a short time. It needs more free electrons to last. When you stopped rubbing the balloon against your hair, free electrons stopped moving to the balloon. That meant that the static electricity stopped too. But if electrons keep moving to the balloon, electricity can last. A steady flow of free electrons is called an electric current.

Electric current moves when free electrons from one atom pass along to the next. To see how this works, stand in a row with your friends. Whisper a word to the person next to you. If everyone whispers the word to the next person, the word reaches the end of the line. Electric current flows along a wire the same way. The wire's atoms stay in place, but the free electrons flow from one atom to the next.

No one moves out of place in your line, but the word you whisper moves from person to person. This is how electricity moves too.

Electric current flows easily through some materials. These materials are called electrical conductors. The metals silver, copper and iron are all good electrical conductors.

This copper wire is a good conductor. It carries electric current well.

The soles of your trainers are made of rubber. Rubber is a very good insulator. Electricity does not travel well through rubber.

Some materials do not carry electric current well. These materials are called insulators. Glass, china, wood, plastic and rubber are good insulators. An insulator can protect you from getting an electric shock. Plastic is wrapped around electrical wiring to stop the electrical current hurting people.

We learned that static electricity does not last long. But it can be very powerful. Did you know that lightning is a kind of static electricity?

You made static electricity when you rubbed a balloon against your hair. But where does an electric current come from? Putting certain materials together can make an electric current. These materials are inside batteries. They make electricity inside the battery.

Batteries come in all shapes and sizes. What have you used batteries for? A torch? A video game?

Batteries are a useful source of electricity. There are many kinds of batteries. We use batteries in torches, MP3 players, CD players, cameras and even cars.

You can use a battery to make an electric current flow. You will need a size D battery, a torch bulb, a clothes peg, tape and a strip of aluminium foil. Fold the foil into a long, thin, flat strip.

How can these things make electricity flow? Find out!

The battery feels cool to the touch. But flowing electricity makes heat.

Look at the battery. The bumpy end of the battery is called the positive terminal. The flat end is the negative terminal. Feel each terminal. Are they warm? No. But you can make them get warm by making an electric current.

Tightly tape one end of the foil strip across the negative terminal. Did the terminal feel warm? No. Now hold the other end of the foil strip across the positive terminal. Are the terminals getting warm now? Yes. That's because electric current is flowing. Electric currents make heat.

The battery should not get warm enough to hurt you. But if it does feel too warm, just put it down!

The loop of foil connects the two terminals. It makes a path from one end of the battery to the other.

The current flowed because you made a path for it to follow. This path is called a circuit. Electricity needs a circuit to flow. The circuit must connect the battery's positive and negative terminals. The circuit must also be closed. A closed circuit is a path without any spaces in it.

Use one hand to make a circle with your thumb and index finger. Can you trace all the way around the circle with the index finger of your other hand? Yes! Your fingers have made a closed circuit.

Holding your fingers in an 'okay' sign can help you think about what a closed circuit looks like. Your fingers make a complete loop.

When you hold your fingers apart, you do not make a complete loop. There is a gap. There is also a gap, or space, in circuits that are not closed.

Now spread your thumb and index finger apart. This time, there is a space between your fingers. You cannot trace a complete circle. Your fingers do not make a closed circuit.

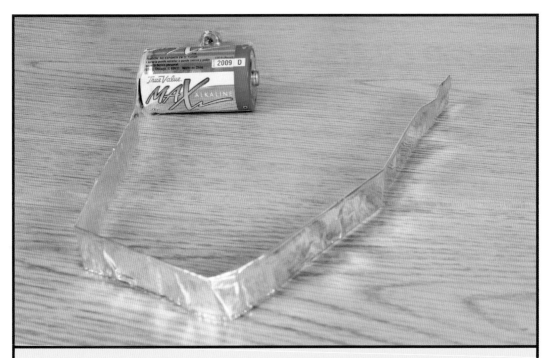

The foil is not making a complete circuit. The loop does not go all the way from the negative terminal to the positive terminal. Electricity cannot travel through the strip.

Look at your battery and the foil strip. One end of the strip is taped to the negative terminal. The other end is not touching the positive terminal. You cannot trace along the foil from one terminal to the other without lifting your finger. It is not a closed circuit. How could you close the circuit?

This circuit is complete. It connects the negative and positive terminals with no gaps in between. Electricity can flow through the foil strip.

Hold the loose end of the foil strip against the positive terminal. Can you trace along the foil from one terminal to the other without lifting your finger? Yes. You have made a closed circuit. You know this because the terminals are warm. That means electrical current is flowing. Current flows only through a closed circuit.

To flow, electric current needs a closed circuit. But what makes the current move? Free electrons need a push to get moving. That push comes from electrical force. Electrical force is measured in units called volts. Your battery has 1.5 volts. That is enough force to push electrical charges along the foil strip.

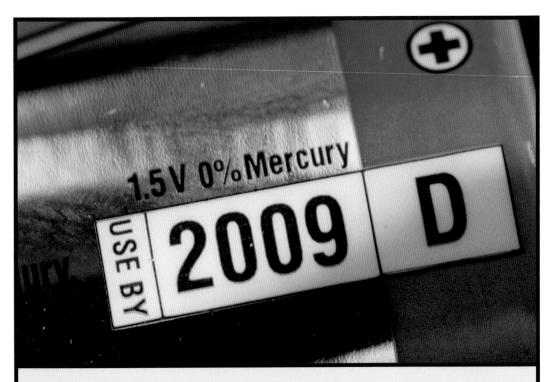

Look for the word *volts* or the letter *V* on your battery. Every battery has a certain number of volts. The more volts a battery has, the more of a 'push' it can give to an electric current.

The foil strip is made of the metal aluminium. Aluminium is an electrical conductor. This means that the strip can carry electricity to the light bulb.

Volts force electric current to keep flowing. Flowing current will light your torch bulb. But to do this, you need a closed circuit. Wrap the loose end of the foil around the metal part of the torch bulb. Clamp the foil in place with the clothes peg. Make sure the other end of the foil is taped to the negative terminal.

The metal point on the bottom of the bulb is called the contact. The metal contact is a good conductor of electricity.

For the bulb to light up, this contact needs to be part of a complete circuit. How can you make a complete circuit?

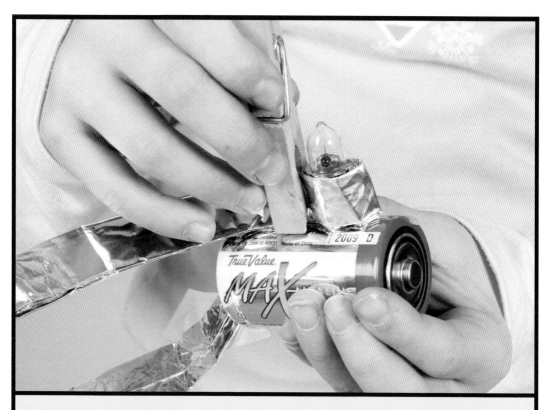

Does the bulb light if you touch the contact to the side of the battery? No, because the battery is covered in plastic. Plastic is an insulator.

Touch the bulb's contact to the negative terminal on the battery. The terminal's metal is a good conductor. Does the bulb light? No. It doesn't light because the negative terminal is not connected to the positive terminal. You have not made a closed circuit yet.

Touch the contact to the positive terminal. Does the bulb light? Yes! Now the circuit is closed. Electric current flows from the negative terminal. It flows through the foil strip to the light bulb and then on to the positive terminal. The circuit is complete. Electric current flows through the wires inside the bulb and makes it light up.

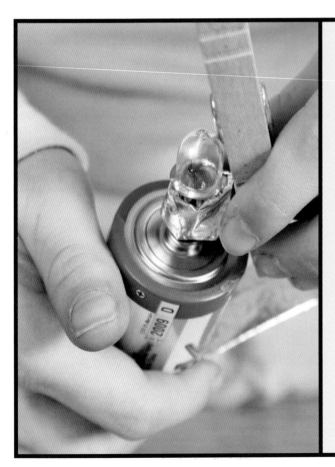

Inside a light bulb are two straight wires connected by a thin, curly wire. When you complete the circuit, the curly wire glows, and the bulb lights up.

Next time you read by a lamp, think about how electricity works!

Electricity is useful for running many kinds of machines and for making heat. It can also make light. And that's very useful for reading a book at night!

ON SHARING A BOOK

When you share a book with a child, you show that reading is important. To get the most out of the experience, read in a comfortable, quiet place. Turn off the television and limit other distractions, such as telephone calls. Be prepared to start slowly. Take turns reading parts of this book. Stop occasionally and discuss what you're reading. Talk about the photographs. If the child begins to lose interest, stop reading. When you pick up the book again, re-read the parts you have already read.

BE A VOCABULARY DETECTIVE

The word list on page 5 contains words that are important in understanding the topic of this book. Be word detectives and search for the words as you read the book together. Talk about what the words mean and how they are used in the sentence. Do any of these words have more than one meaning? You will find the words defined in a glossary on page 46.

WHAT ABOUT QUESTIONS?

Use questions to make sure the child understands the information in this book. Here are some suggestions:

> What did this paragraph tell us? What does this picture show? What do you think we'll learn about next? What are the three kinds of parts in an atom? Do electrons have a positive or a negative charge? What is static electricity? What do we call a material that carries electric current well? What is a closed circuit? What is your favourite part of the book? Why?

If the child has questions, don't hesitate to respond with questions of your own, such as What do *you* think? Why? What is it that you don't know? If the child can't remember certain facts, turn to the index.

INTRODUCING THE INDEX

The index helps readers find information without searching through the whole book. Turn to the index on page 48. Choose an entry, such as *circuits,* and ask the child to use the index to find out the differences between a closed and an open circuit. Repeat with as many entries as you like. Ask the child to point out the differences between an index and a glossary. (The index helps readers find information, while the glossary tells readers what words mean.)

LEARN MORE ABOUT
ELECTRICITY AND ENERGY

BOOKS

Bailey, Jacqui. *How Do We Use Electricity?* (Investigating Science) Franklin Watts, 2006.

Ballard, Carol. *Electricity* (How Does Science Work?) Wayland, 2006.

Hunter, Rebecca. *The Facts About Electricity* (Science the Facts) Franklin Watts, 2003.

Llewellyn, Claire. *Electricity* (Start-Up Science) Evans Brothers Ltd, 2004.

Parker, Steve. *Electricity* Dorling Kindersley, 2000.

WEBSITES

Dialogue for Kids: Electricity
http://idahoptv.org/dialogue4kids/season3/electricity/facts.html
This website talks about where electricity comes from, types of electricity, how it is used and more.

Energy Kid's Page: Energy History
http://www.eia.doe.gov/kids/history/index.html
This website from the Energy Information Administration has timelines exploring important discoveries about energy.

NASA Science Files: Understanding Electricity
http://whyfiles.larc.nasa.gov/text/kids/Problem_Board/problems /electricity/electricity2.html
This site presents an overview of basic ideas about electricity.

BBC Schools - Pod's Mission.
http://www.bbc.co.uk/schools/podsmission/electricity/
This site helps children to learn the basic principles of electrical circuits with the game Pod's mission.

GLOSSARY

atoms: the tiny particles that make up all things

circuit: the path that an electric current follows

conductors: materials that carry electric current well

current: the flow of electricity through something

electrical charge: the energy of an atom or part of an atom. When atoms gain or lose electrons, they gain electrical charge.

electrons: the parts of an atom that have a negative charge. Electrons circle around the centre of the atom.

insulator: a material that does not carry electric current well

ion: an atom that has gained or lost electrons

negative charge: the charge that a substance has if its atoms have gained extra electrons from other atoms

nucleus: the centre of an atom. The nucleus is made of protons and neutrons.

orbit: a circular or oval path. Electrons follow an orbit around the centre of an atom.

particles: tiny pieces

positive charge: the charge that a substance has if its atoms have lost electrons to other atoms

static electricity: energy created between objects that have different electric charges

terminal: one of the ends of a battery. Every battery has a positive terminal and a negative terminal.

INDEX

Pages listed in **bold** type refer to photographs.

atoms 10–15

batteries 28–**29, 30, 31–33, 36–39, 41–42**

circuits 33–37, 38, 39, 41, 42
conductors **26**, 39, 40, 41
current 24–27, 28, 30, 31, 32–33, 37, 38, 39, 42

electrical charge 15, 16,
electrons 12, 13, 14–15, 16, 24–25

insulators **27**, 41
ions 15

lightning **8, 28**

negative charge 12, 15, 18, 21, 23
nucleus **12**, 13

orbit **12**, 13, 14, 18

particles 10
plastic 27
positive charge 12, 15

static electricity 20–21, 24, 28

terminal 31-33, 36, 37, 41, 42

volts 38